PANAMA BIRDS

WESTERN PANAMA

...ONS

Mourning Warbler
Reinita enlutada
Oporornis philadelphia 13 cm

Bananaquit
Mielero reinita
Coereba flaveola 10 cm

Slate-throated Redstart
Candelita gargantiplomiza
Myioborus miniatus 13 cm

Canada Warbler
Reinita collareja
Wilsonia canadensis 12 cm

Red-throated Ant-Tanager
Tangara hormiguera gorguirroja
Habia fuscicauda 18 cm

Common Bush-Tanager
Tangara de monte común
Chlorospingus ophthalmicus 13 cm

Passerini's Tanager (Atlantic) ♀
Tangara de Passerini
Ramphocelus passerinii 15 cm
Cherrie's Tanager (Pacific)
Tangara de Cherrie
Ramphocelus costaricensis 15 cm

Crimson-backed Tanager
Tangara dorsirroja
Ramphocelus dimidiatus 15 cm

Silver-throated Tanager
Tangara goliplata
Tangara icterocephala 13 cm

White-shouldered Tanager
Tangara hombriblanca
Tachyphonus luctuosus 13 cm

Golden-hooded Tanager
Tangara capuchidorada
Tangara larvata 13 cm

Speckled Tanager
Tangara moteada
Tangara guttata 13 cm

Summer Tananger
Tangara veranera
Piranga rubra 18 cm

Bay-headed Tanager
Tangara cabecibaya
Tangara gyrola 13 cm

Blue-gray Tanager
Tangara azuleja
Thraupis episcopus 15 cm

Palm Tanager
Tangara palmera
Thraupis palmarum 15 cm

Green Honeycreeper
Mielero verde
Chlorophanes spiza 13 cm

Blue Dacnis
Dacnis azul
Dacnis cayana 10 cm

Scarlet-thighed Dacnis
Dacnis musliescarlata
Dacnis venusta 13 cm

Red-legged Honeycreeper
Mielero patirrojo
Cyanerpes cyaneus 10 cm

White-collared Seedeater
Espiguero cueliblanco
Sporophila torqueola 10 cm

Ruddy-breasted Seedeater
Espiguero pechirrojizo
Sporophilia minuta 10 cm

Variable Seedeater
Espiguero variable
Sporophila americana 10 cm

Yellow-bellied Seedeater
Espiguero ventriamarillo
Sporophila nigricollis 10 cm

Yellow-faced Grassquit
Semillerito cariamarillo
Tiaris olivaceus 10 cm

Blue-black Grassquit
Semillerito negriazulado
Volatinia jacarina 10 cm

Gray-chested Dove
Paloma pechigris
Leptotila cassinii 24 cm

White-tipped Dove
Paloma rabiblanca
Leptotila verreauxi 26 cm

Morning Dove
Tórtola rabiaguda
Zenaida macroura 30 cm

Ruddy Ground-Dove
Tortolita rojiza
Columbina talpacoti 16.5 cm

Sulphur-winged Parakeet
Perico aliamarillo
Pyrrhura hoffmanni 23 cm

Olive-throated Parakeet
Perico azteca
Aratinga nana 23 cm

Brown-throated Parakeet
Perico carisucio
Aratinga pertinax 24 cm

Orange-chinned Parakeet
Periquito barbinaranja
Brotogeris jugularis 17 cm

Brown-hooded Parrot
Loro cabecipardo
Pionopsitta haematotis 23 cm

Crimson-fronted Parakeet
Perico frentirrojo
Aratinga finschi 28 cm

Red-lored Amazon
Loro frentirrojo
Amazona autumnalis 34 cm

Blue-headed Parrot
Loro cabeciazul
Pionus menstruus 25 cm

Yellow-crowned Amazon
Lora coroniamarillo
Amazona orchrocephala 30 cm

Mealy Amazon
Loro harinoso
Amazona farinosa 38 cm

Striped Cuckoo
Cuclillo listado
Tapera naevia 30 cm

Squirrel Cuckoo
Cuco ardilla
Piaya cayana 46 cm

Groove-billed Ani
Carrapatero piquiestriado
Crotophaga sulcirostris 30 cm

Common Pauraque
Tapacamino común
Nyctidromus albicollis 28 cm

Mottled Owl
Búho moteado
Ciccaba virgata 35 cm

Smooth-billed Ani
Garrapatero piquiliso
Crotophaga ani 36 cm

Lesser Nighthawk
Añapero menor
Chordeiles acutipennis 23 cm

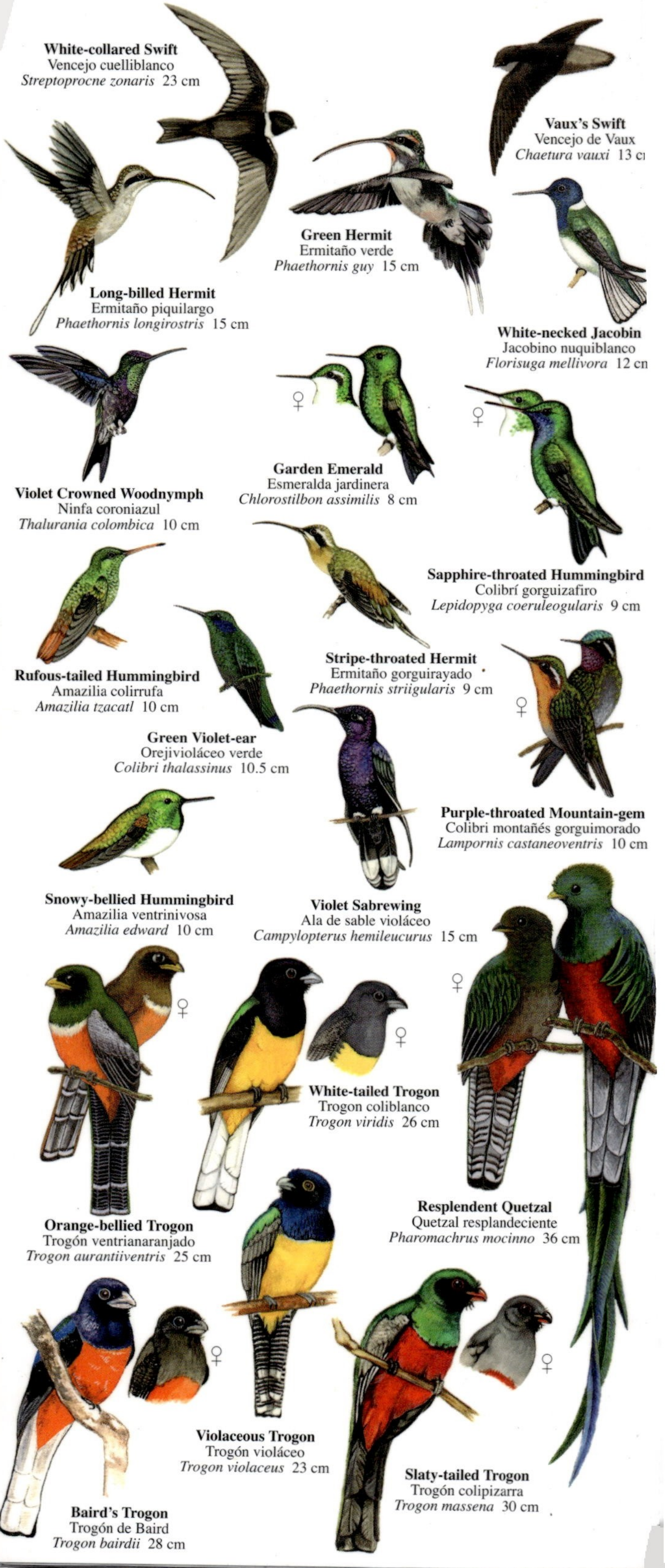
White-collared Swift
Vencejo cuelliblanco
Streptoprocne zonaris 23 cm
Vaux's Swift
Vencejo de Vaux
Chaetura vauxi 13 c
Long-billed Hermit
Ermitaño piquilargo
Phaethornis longirostris 15 cm
Green Hermit
Ermitaño verde
Phaethornis guy 15 cm
White-necked Jacobin
Jacobino nuquiblanco
Florisuga mellivora 12 c
♀
Garden Emerald
Esmeralda jardinera
Chlorostilbon assimilis 8 cm
♀
Violet Crowned Woodnymph
Ninfa coroniazul
Thalurania colombica 10 cm
Sapphire-throated Hummingbird
Colibrí gorguizafiro
Lepidopyga coeruleogularis 9 cm
Rufous-tailed Hummingbird
Amazilia colirrufa
Amazilia tzacatl 10 cm
Stripe-throated Hermit
Ermitaño gorguirayado
Phaethornis striigularis 9 cm
♀
Green Violet-ear
Orejivioláceo verde
Colibri thalassinus 10.5 cm
Purple-throated Mountain-gem
Colibri montañés gorguimorado
Lampornis castaneoventris 10 cm
Snowy-bellied Hummingbird
Amazilia ventrinivosa
Amazilia edward 10 cm
Violet Sabrewing
Ala de sable violáceo
Campylopterus hemileucurus 15 cm
♀
♀
♀
White-tailed Trogon
Trogon coliblanco
Trogon viridis 26 cm
Orange-bellied Trogon
Trogón ventrianaranjado
Trogon aurantiiventris 25 cm
Resplendent Quetzal
Quetzal resplandeciente
Pharomachrus mocinno 36 cm
♀
♀
Violaceous Trogon
Trogón violáceo
Trogon violaceus 23 cm
Slaty-tailed Trogon
Trogón colipizarra
Trogon massena 30 cm
Baird's Trogon
Trogón de Baird
Trogon bairdii 28 cm

Rufous Motmot
Momoto rufo
Baryphthengus martii 46 cm

Broad-billed Motmot
Momoto piquiancho
Electron platyrhynchum 30 cm

Blue-crowned Motmot
Momoto coroniazulado
Momotus momota 41 cm

Ringed Kingfisher
Martín pescador grande
Ceryle torquatus 41 cm

Green Kingfisher
Martín pescador verde
Chloroceryle americana 18 cm

Belted Kingfisher
Martín pescador norteño
Ceryle alcyon 30 cm

Keel-billed Toucan
Tucán pico iris
Ramphastos sulfuratus 46 cm

Chestnut-mandibled Toucan
Tucán de Swainson
Ramphastos swainsonii 56 cm

Collared Aracari
Tucancillo collarejo
Pteroglossus torquatus 41 cm

Red-crowned Woodpecker
Carpintero coronirrojo
Melanerpes rubricapillus 18 cm

♀

Acorn Woodpecker
Carpintero careto
Melanerpes formicivorus 23 cm

♀

Lineated Woodpecker
Carpintero lineado
Dryocopus lineatus 33 cm

Black-cheeked Woodpecker
Carpintero carinegro
Melanerpes pucherani 18 cm

Golden-olive Woodpecker
Carpintero olividorado
Piculus rubiginosus 20 cm

♀

Pale-breasted Spinetail
Colaespina pechiblanca
Synallaxis albescens 15 cm

Red-faced Spinetail
Colaespina carirroja
Cranioleuca dissita 15 cm

Lineated Foliage-gleaner
Limpiafronda lineada
Syndactyla subalaris 18 cm

Spotted Barbtail
Subepalo moteado
Premnoplex brunnescens 15 cm

Black-necked Stilt
Cigüeñuela cuellinegra
Himantopus mexicanus 38 cm

Northern Jacana
Jacana norteña
Jacana spinosa 23 cm

♀

Wattled Jacana
Jacana carunculada
Jacana jacana 23 cm

Lesser Yellowlegs
Patiamarillo menor
Tringa flavipes 28 cm

Willet
Playero aliblanco
Tringa semipalmata 38 cm

Greater Yellowlegs
Patiamarillo mayor
Tringa melanoleuca 36 cm

Ruddy Turnstone
Vuelvepiedras rojizo
Arenaria interpres 23 cm

Western Sandpiper
Playero occidental
Calidris mauri 15 cm

Whimbrel
Zarapito trinador
Numenius phaeopus 43 cm

Spotted Sandpiper
Playero coleador
Actitis macularia 20 cm

Semipalmated Sandpiper
Playero semipalmeado
Calidris pusilla 15 cm

breeding ♀

non breeding

Red-necked Phalarope
Falaropo picofino
Phalaropus lobatus 20 cm

Least Sandpiper
Playero menudo
Calidris minutilla 15 cm

Laughing Gull
Gaviota reidora
Larus atricilla 41 cm

Short-billed Dowitcher
Agujeta piquicorta
Limnodromus griseus 27 cm

Royal Tern
Gaviotín real
Sterna maxima 51 cm

Sandwich Tern
Gaviotín puntiamarillo
Sterna sandvicensis 41 cm

Ruddy Pigeon
Paloma rojiza
Patagioenas subvinacea 28 cm

Pale-vented Pigeon
Paloma colorada
Patagioenas cayennensis 30 cm

Band-tailed Pigeon
Paloma collareja
Patagioenas fasciata 36 cm

Short-billed Pigeon
Paloma piquicorta
Patagioenas nigrirostris 26.5 cm

Scaled Pigeon
Paloma escamosa
Patagioenas speciosa 32 cm

Mangrove Black-Hawk
Gavilán manglero
Buteogallus subtilis 56 cm

Common Black-Hawk
Gavilán cangrejero
Buteogallus anthracinus 56 cm

White Hawk
Gavilán blanco
Leucopternis albicollis 58 cm

Broad-winged Hawk
Gavilán aludo
Buteo platypterus 41 cm

Barred Forest-Falcon
Halcón montés barreteado
Micrastur ruficollis 38 cm

Collared Forest Falcon
Halcón montés collarejo
Micrastur semitorquatus 61 cm

Black Hawk-Eagle
Aguililo blanquinegro
Spizaestus tyrannus 66 cm

Roadside Hawk
Gavilán caminero
Buteo magnirostris 38 cm

Yellow-headed Caracara
Caracara cabeciamarilla
Milvago chimachima 41 cm

Crested Carcara
Carcara crestada
Caracara cheriway 58 cm

Gray-necked Wood-Rail
Rascón montés cuelligris
Aramides cajanea 38 cm

Purple Gallinule
Gallareta morada
Porphyrio martinica 33 cm

White-throated Crake
Polluela gargantiblanca
Laterallus albigularis 15 cm

Common Moorehen
Gallareta frentirroja
Gallinula chloropus 36 cm

Black-bellied Plover
Chorlo gris
Pluvialis squatarola 30 cm

Collared Plover
Chorlo collarejo
Charadrius collaris 15 cm

Semipalmated Plover
Chorlo semipalmeado
Charadrius semipalmatus 18 cm

Spotted Woodcreeper
Trepatroncos manchado
Xiphorhynchus erythropygius 23 cm

Cocoa Woodcreeper
Trepatroncos chocolate
Xiphorhynchus susurrans 23 cm

Plain Xenops
Xenops bayo
Xenops minutus 13 cm

Wedge-billed Woodcreeper
Trepatroncos pico de cuña
Glyphorynchus spirurus 15 cm

♀

Barred Antshrike
Batará barreteado
Thamnophilus doliatus 16 cm

♀

Western Slaty-Antshrike
Batará pizarroso occidental
Thamnophilus atrinucha 15 cm

♀

Fasciated Antshrike
Batará lineado
Cymbilaimus lineatus 18 cm

Russet Antshrike
Batará rufo
Thamnistes anabatinus 15 cm

♀

Dot-winged Antwren
Hormiguerito alipunteado
Microrhopias quixensis 10 cm

♀

Checker-throated Antwren
Hormiguerito leonado
Myrmotherula fulviventris 10 cm

♀

Dusky Antbird
Hormiguero negruzco
Cercomacra tyrannina 15 cm

Chestnut-backed Antbird
Hormiguero dorsicastaño
Myrmeciza exsul 14 cm

♀

White-flanked Antwren
Hormiguerito flanquiblanco
Myrmotherula axillaris 10 cm

Black-faced Antthrush
Formicario carinegro
Formicarius analis 18 cm

♀

Spotted Antbird
Hormiguero collarejo
Hylophylax naevioides 10 cm

Southern Beardless-Tyrannulet
Tiranolete silbador sureño
Camptostoma obsoletum 10 cm

Torrent Tyrannulet
Mosquerito guardarríos
Serpophaga cinerea 10 cm

Yellow-crowned Tyrannulet
Tiranolete coroniamarillo
Tyrannulus elatus 10 cm

Paltry Tyrannulet
Tiranolete cejigris
Zimmerius vilissimus 10 cm

Yellow-bellied Elaenia
Elenia penachuda
Elaenia flavogaster 15 cm

Ochre-bellied Flycatcher
Mosquerito ventriocráceo
Mionectes oleagineus 13 cm

Common Tody-Flycatcher
Espatulilla común
Todirostrum cinereum 10 cm

Scale-crested Pygmy-Tyrant
Tirano-enano crestiescamado
Lophotriccus pileatus 9 cm

Olive-sided Flycatcher
Pibí boreal
Contopus cooperi 17 cm

Mountain Elaenia
Elenia montañera
Elaenia frantzii 15 cm

Eastern Wood Pewee
Pibí oriental
Contopus virens 15 cm

Dusky-capped Flycatcher
Copetón crestioscuro
Myiarchus tuberculifer 15 cm

Great Crested Flycatcher
Copetón viajero
Myiarchus crinitus 20 cm

Great Kiskadee
Bienteveo grande
Pitangus sulphuratus 23 cm

Boat-billed Flycatcher
Mosquero picudo
Megarynchus pitangua 23 cm

Social Flycatcher
Mosquero social
Myiozetetes similis 15 cn

Piratic Flycatcher
Mosquero pirata
Legatus leucophaius 15 cr

Bright-rumped Attila
Atila lomiamarilla
Attila spadiceus 18 cm

Streaked Flycatcher
Mosquero rayado
Myiodynastes maculatus 20 cm

Tropical Kingbird
Tirano tropical
Tyrannus melancholicus 21 cm

Panama Flycatcher
Copetón panameño
Myiarchus panamensis 20 cm

Eastern Kingbird
Tirano norteño
Tyrannus tyrannus 20 cm

Black Phoebe
Mosquerito negro
Sayornis nigricans 18 cm

Fork-tailed Flycatcher
Tijereta sabanera
Tyrannus savana 36 cm

Masked Tityra
Titira enmascarada
Tityra semifasciata 21 cm

Three-wattled Bellbird
Campanero tricarunculado
Procnias tricarunculatus 25 cm

Lance-tailed Manakin
Saltarín coludo
Chiroxiphia lanceolata 15 cm

Golden-collared Manakin
Saltarín cuellidorado
Manacus vitellinus 11 cm

Blue-crowned Manakin
Saltarín coroniceleste
Pipra coronata 10 cm

Yellow-green Vireo
Vireo verdiamarillo
Vireo flavoviridis 15 cm

Red-capped Manakin
Saltarín cabecirrojo
Pipra mentalis 10 cm

Scrub Greenlet
Verdillo matorralero
Hylophilus flavipes 13 cm

Red-eyed Vireo
Vireo ojirrojo
Vireo olivaceus 15 cm

ellow-throated Vireo
Vireo pechiamarillo
ireo flavifrons 13 cm

Black-cheasted Jay
Urraca pechinegra
Cyanocorax affinis 33 cm

Lesser Greenlet
Verdillo menor
Hylophilus decurtatus 10 cm

Gray-breasted Martin
Martín pechigris
Progne chalybea 18 cm

Little Blue Heron
Garza azul chica
Egretta caerulea 61 cm

Tricolored Heron
Garza tricolor
Egretta tricolor 66 cm

Yellow-crowned Night-Heron
Garza nocturna cabeciamarilla
Nyctanassa violacea 61 cm

Green Heron
Garza verde
Butorides virescens 48 cm

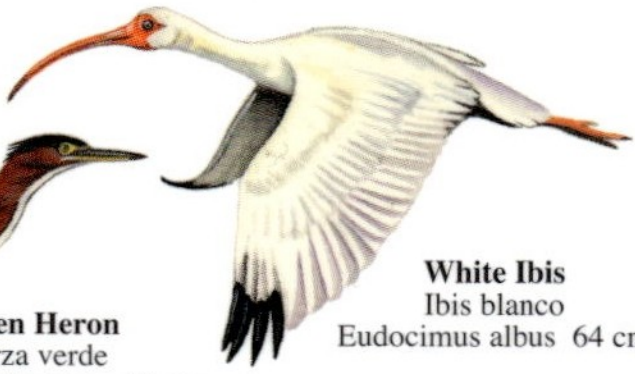

White Ibis
Ibis blanco
Eudocimus albus 64 cm

Boat-billed Heron
Garza cucharón
Cochlearius cochlearius 51 cm

Glossy Ibis
Ibis morito
Plegadis falcinellus 58 cm

Turkey Vulture
Gallinazo cabecirrojo
Cathartes aura 76 cm

Black Vulture
Gallinazo negro
Coragyps atratus 64 cm

Lesser Yellow-headed Vulture
Gallinazo cabeciamarillo menor
Cathartes burrovianus 64 cm

Gray-headed Kite
Elanio cabecigris
Leptodon cayanensis 51 cm

Swallow-tailed Kite
Elanio tijereta
Elanoides forficatus 58 cm

Osprey
Aguila pescadora
Pandion haliaetus 58 cm

Double-toothed Kite
Elanio bidentado
Harpagus bidentatus 36 cm

White-tailed Kite
Elanio coliblanco
Elanus leucurus 41 cm

Swainson's Hawk
Gavilán de Swainson
Buteo swainsoni 53 cm

Great Tinamou
Tinamú grande
Tinamus major 43 cm

Little Tinamou
Tinamú chico
Crypturellus soui 23 cm

Black-bellied Whistling-Duc
Pato silbador aliblanco
Dendrocygna autumnalis 53 c

Blue-winged Teal
Cerceta aliazul
Anas discors 38 cm

Gray-headed Chachalaca
Chachalaca cabeigris
Ortalis cinereiceps 51 cm

Least Grebe
Zambullidor menor
Tachybaptus dominicus 23 cm

Pied-billed Grebe
Zambullidor piquipinto
Podilymbus podiceps 33 cm

Brown Pelican
Pelican pardo, Old Joe
Pelecanus occidentalis 116 cm

Blue-footed Booby
Piquero patiazul
Sula nebouxii 81 cm

Brown Booby
Piquero pardo
Sula leucogaster 70 cm

Neotropic Cormorant
Cormorán neotropical
Phalacrocorax brasilianus 66 cm

Magnificent Frigatebird
Fragata magnífica
Fregata magnificens 102 cm

Anhinga
Aninga
Anhinga anhinga 89 cm

Snowy Egret
Garza nívea
Egretta thula 61 cm

Great Egret
Garceta grande
Ardea alba 99 cm

Cattle Egret
Garceta bueyera
Bubulcus ibis 51 cm

Great Blue Heron
Garza azul mayor
Ardea herodias 117 cm

Southern Rough-winged Swallow
Golondrina alirrasposa sureña
Stelgidopteryx ruficollis 13 cm

Blue-and-white Swallow
Golondrina azul y blanca
Pygochelidon cyanoleuca 10 cm

Mangrove Swallow
Golondrina manglera
Tachycineta albilinea 13 cm

Rufous-breasted Wren
Sotorey pechirrufo
Thryothorus rutilus 13 cm

Northern Rough-winged Swallow
Golondrina alirrasposa norteña
Stelgidopteryx serripennis 13 cm

Barn Swallow
Golondrina tijereta
Hirundo rustica 13 cm

Black-bellied Wren
Sotorrey ventrinegro
Tryothorus fasciatoventris 15 cm

Gray-breasted Wood-Wren
Sotorrey selvático pechigris
Henicorhina leucophrys 10 cm

Plain Wren
Sotorrey modesto
Thryothorus modestus 13 cm

House Wren
Soterrey común
Troglodytes aedon 10 cm

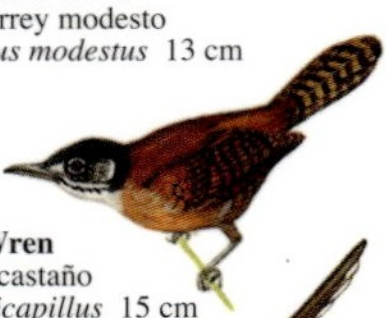

Bay Wren
Sotorrey castaño
Thryothorus nigricapillus 15 cm

White-breasted Wood-Wren
Sotorrey selvático pechiblanco
Henicorhina leucosticta 10 cm

Long-billed Gnatwren
Soterillo piquilargo
Ramphocaenus melanurus 11 cm

♀

Tropical Gnatcatcher
Perlita tropical
Polioptila plumbea 10 cm

Southern Nightingale-Wren
Sotorrey-ruiseñor sureño
Microcerculus marginatus 10 cm

Orange-billed Nightingale-Thrush
Zorzal piquinaranja
Catharus aurantiirostris 15 cm

Slaty-backed Nightingale-Thrush
Zorzal sombrío
Catharus fuscater 18 cm

♀

Black-faced Solitaire
Solitario carinegro
Myadestes melanops 18 cm

Swainson's Thrush
Zorzal de Swainson
Catharus ustulatus 18 cm

Clay-colored Thrush
Mirlo pardo
Turdus grayi 23 cm

♀

Black-and-white Warbler
Reinita trepadora
Mniotilta varia 13 cm

Tennessee Warbler
Reinita verdilla
Oreothlypis peregrina 13 cm

Prothonotary Warbler
Reinita protonotaria
Protonotaria citrea 13 cm

Yellow Warbler
Reinita amarilla
Dendroica petechia 13 cm

Chestnut-sided Warbler
Reinita flanquicastaña
Dendroica pensylvanica 13 cm

Northern Waterthrush
Reinita acuática norteña
Parkesia noveboracensis 15 cm

Slaty Flowerpiecer
Pinchaflor pizarroso
Diglossa plumbea 10 cm

Chestnut-capped Brush-Finch
Matorralero gorricastaño
Buarremon brunneinucha 18 cm

White-napped Brush-Finch
Matorralero nuquiblanco
Atlapetes albinucha 18 cm

Orange-billed Sparrow
Gorrión piquinaranja
Arremon aurantiirostris 15 cm

Rufous-collared Sparrow
Gorrión ruficollarejo
Zonotrichia capensis 13 cm

Black-striped Sparrow
Gorrión negrilistado
Arremonops conirostris 15 cm

House Sparrow
Gorrión doméstica
Passer domesticus 15 cm

Rose-breasted Grosbeak
Picogrueso pechirrosado
Pheucticus ludovicianus 20 cm

Red-breasted Blackbird
Pastorero pechirrojo
Sturnella militaris 18 cm

Eastern Meadowlark
Pastorero oriental
Sturnella magna 20 cm

Blue-black Grosbeak
Picogrueso negriazulado
Cyanocompsa cyanoides 18 cm

Orchard Oriole
Bolsero castaño
Icterus spurius 18 cm

Great-tailed Grackle
Negro coligrande
Quiscalus mexicanus 43 cm

Bronzed Cowbird
Vaquero ojirrojo
Molothus aeneus 20 cm

Baltimore Oriole
Bolsero de Baltimore
Icterus galbula 20 cm

Buff-throated Saltator
Saltator gorguianteado
Saltator maximus 20 cm

Scarlet-rumped Cacique
Cacique lomiescarlata
Cacicus uropygialis 23 cm

Yellow-backed Oriole
Bolsero dorsiamarillo
Icterus chrysater 22 cm

Yellow-rumped Cacique
Cacique lomiamarillo
Cacicus cela 28 cm

Chestnut-headed Oropendola
Oropéndola cabecicastaña
Psarocolius wagleri 36 cm

Lesser Goldfinch
Jilguero menor
Spinus psaltria 10 cm

Montezuma Oropendola
Oropéndola de Montezuma
Psarocolius montezuma 51 cm

Yellow-crowned Euphonia
Eufonia coroniamarilla
Euphonia luteicapilla 10 cm

Thick-billed Euphonia
Eufonia piquigruesa
Euphonia laniirostris 10 cm

There have been 894 different species of birds recorded in Panama. This guide shows 259 of the most commonly seen birds in Western Panama.

Se han identificado 894 especies diferentes de aves en Panamá. Esta guía muestra 259 de las especies más comunes observadas en la zona oeste de Panamá.

HARPY EAGLE
NATIONAL BIRD

The Harpy Eagle (*Harpia harpyja*) is the national bird of the Republic of Panama. It has a body length up to 104 cm (41 inches) and a wing span to 213 cm (7 feet). The Harpy is the strongest bird of prey in the world. They feed primarily on animals that live in trees like sloths, monkeys and opossums. They have no natural predators but have disappeared from most of their historical range (Mexico to Argentina) mainly due to habitat loss and persecution.

El águila arpía (*Harpia harpyja*) es el ave nacional de Panamá. El cuerpo llega a medir hasta 104 cm (41 pulgadas) y una envergadura alar de 213 cm (7 feet). El águila arpía es el ave más fuerte de las aves de presa. Se alimentan principalmente de animales que viven en los árboles como perezosos, monos y zorros. No tienen depredadores naturales, pero han desaparecido de la mayor parte de su hábitat histórico (México a Argentina), principalmente debido a la pérdida de hábitat y persecución.

Western Panama includes the provinces of Bocas del Toro, Chiriquí, Veraguas, Herrera and Los Santos.

Panamá oeste abarca las provincias de Bocas del Toro, Chiriquí, Veraguas, Herrera y Los Santos.

Rainforest Publications gives a donation for the protection of Tropical Forests with each guide that is purchased. Log on to our web site to see where the donations go.

Birds illustrated by Robert Dean and Mark Wainwright.

Printed in the US on recycled paper.

www.rainforestpublications.com

ISBN 1888538-72-4